21st Century
Basic Skills
Library

GRAPHING STORY PROBLEMS

My Birthday Presents

From Mom

From Dad

From Sis

by Sherra G. Edgar

Cherry Lake Publishing • Ann Arbor, Michigan

2

Published in the United States of America by Cherry Lake Publishing
Ann Arbor, Michigan
www.cherrylakepublishing.com

Consultants: Janice Bradley, PhD, Mathematically Connected Communities, New Mexico State University; Marla Conn, Read-Ability
Editorial direction and book production: Red Line Editorial

Photo Credits: Shutterstock Images, cover, 1, 6, 10 (left), 10 (middle), 10 (right), 20; iStockphoto/Thinkstock, 4, 16; Monkey Business/Thinkstock, 8; Jupiter Images/Thinkstock, 12

Library of Congress Cataloging-in-Publication Data
Edgar, Sherra G.
  Graphing story problems / Sherra G. Edgar.
      pages cm. -- (Let's make graphs)
  Audience: 6.
  Audience: K to grade 3.
  Includes bibliographical references and index.
  ISBN 978-1-62431-391-2 (hardcover) -- ISBN 978-1-62431-467-4 (pbk.) -- ISBN 978-1-62431-429-2 (pdf) -- ISBN 978-1-62431-505-3 (ebook)
  1. Graphic methods--Juvenile literature. 2. Mathematical analysis--Juvenile literature. I. Title.

QA90.E344 2014
001.4'226--dc23

2013005344

Cherry Lake Publishing would like to acknowledge the work of The Partnership for 21st Century Skills. Please visit www.p21.org for more information.

Printed in the United States of America
Corporate Graphics Inc.
July 2013
CLFA11

TABLE OF CONTENTS

Flowers

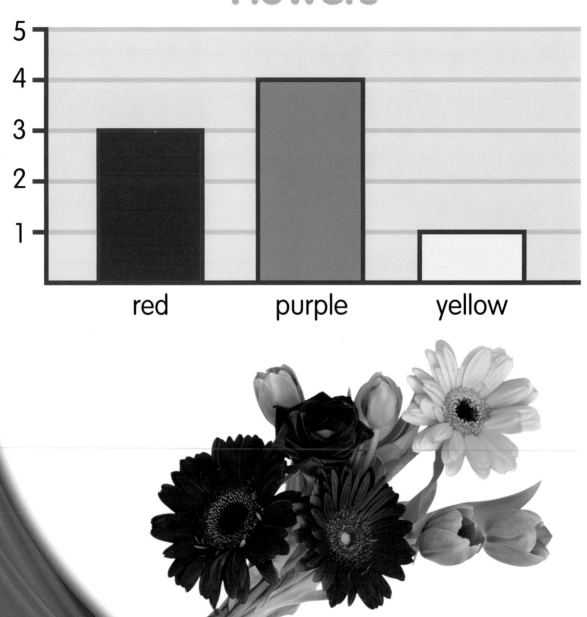

red	purple	yellow

Graphs Solve Problems

We use **graphs** to **compare data**. On a **bar graph, bars** stand for **amounts**.

You can make a bar graph to solve problems.

Using a Graph

Jill is bringing snacks tomorrow. She wants to bring snacks people like.

How can Jill make sure she brings the right snack? Jill will use a bar graph.

Jill asked kids which snack they would like. Then she counted their answers.

Snacks

10
9
8
7
6
5
4
3
2
1

Jill made a big **L**. She numbered 1 to 10 up the side of the **L**.

Snacks

10			
9			
8			
7			
6			
5			
4			
3			
2			
1			
	oranges	carrots	crackers

Jill wrote **labels** under the **L**: oranges, carrots, and crackers. She made a bar for each snack.

Snacks

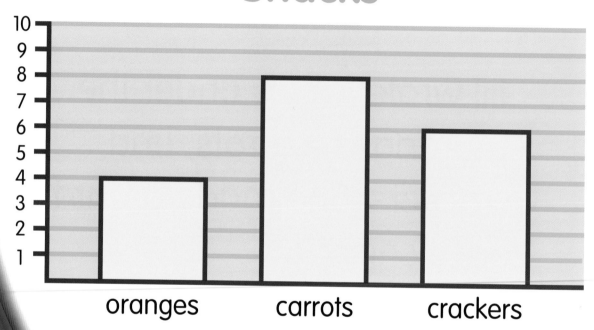

What does Jill's graph show?
Most of Jill's friends like
carrots. Jill will bring carrots
for a snack.

You Try It!

Jake wants to know which park has the most slides. How can he use a graph to solve his problem?

Find Out More

BOOK

Roy, Jennifer Rozines, and Gregory Roy. *Graphing in the Desert*. New York: Marshall Cavendish, 2007.

WEB SITE

Illuminations Data Grapher
http://illuminations.nctm.org/ActivityDetail.aspx?ID=204
Type in your data and this web site will create a graph for you.

Glossary

amounts (uh-MOUNTS) how many or how much there is of something

bars (BAHRS) solid rectangles that stand for numbers on a bar graph

bar graph (bahr graf) a graph that uses bars to stand for numbers

compare (kuhm-PAIR) to show how things are alike

data (DEY-tah) amounts from a graph

graphs (GRAFS) pictures that compare two or more amounts

labels (LEY-buhls) names

Home and School Connection

Use this list of words from the book to help your child become a better reader. Word games and writing activities can help beginning readers reinforce literacy skills.

a	graph	park	under
amounts	graphs	people	up
answers	has	problem	use
asked	he	problems	wants
bar	his	right	we
bars	how	she	what
big	is	show	which
bring	Jake	side	will
bringing	Jill	slides	would
brings	kids	snack	wrote
can	know	snacks	you
carrots	labels	solve	
compare	like	stand	
counted	made	sure	
crackers	make	the	
data	most	their	
does	numbered	then	
each	of	they	
for	on	to	
friends	oranges	tomorrow	

Index

About the Author

Sherra G. Edgar is a former primary school teacher who now writes books for children. She also writes a blog for women. She lives in Texas with her husband and son. She loves reading, writing, and spending time with friends and family.